JKJC

THE STATUE OF
LIBERTY

Ann Tatlock

PURPLE TOAD
PUBLISHING

PURPLE TOAD
PUBLISHING

Printing   1   2   3   4   5   6   7   8   9

Big Ben
The Eiffel Tower
The Space Needle
The Statue of Liberty
The Sydney Opera House
The Taj Mahal

Publisher's Cataloging-in-Publication Data
Tatlock, Ann.
  Statue of Liberty / written by Ann Tatlock.
     p. cm.
Includes bibliographic references, glossary, and index.
ISBN 9781624692079
1. Statue of Liberty National Monument (N.Y. and N.J.)—Juvenile literature. 2. Architecture—Vocational guidance—Juvenile literature. I. Series: Building on a Dream.
 NA2555 2017
 507.8
                                    Library of Congress Control Number: 2016937176

eBook ISBN: 9781624692086

ABOUT THE AUTHOR: Ann Tatlock is a novelist and children's book author. Her works have received numerous awards, including the Silver Angel Award from Excellence in Media and the Midwest Book Award. She lives in the Blue Ridge Mountains of western North Carolina with her husband and daughter.

CONTENTS

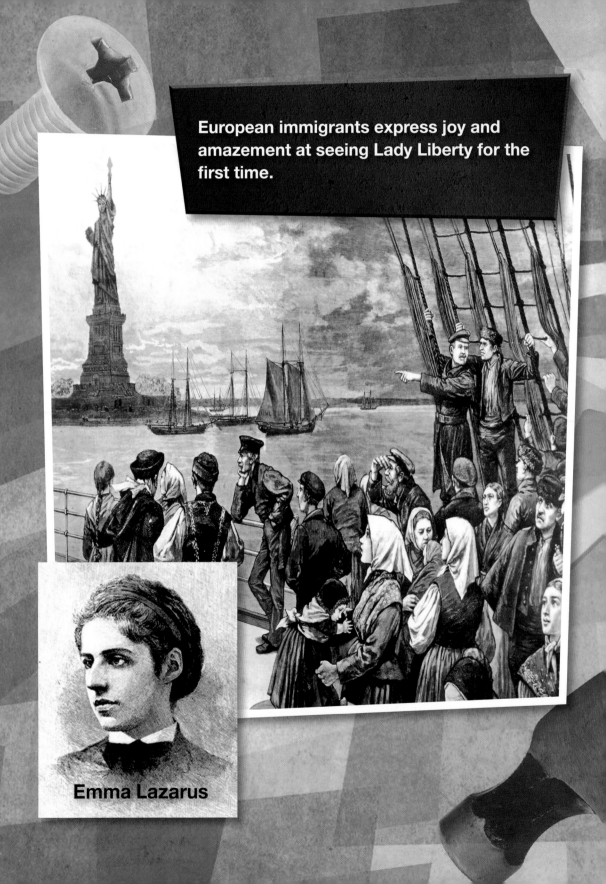

European immigrants express joy and amazement at seeing Lady Liberty for the first time.

Emma Lazarus

The Heart of Lady Liberty

For a long time, immigrants have been coming to the United States to begin new lives. Many arrived by sea, sailing into New York Harbor. Since 1886, the Statue of Liberty has been there to greet them.

To those about to make America their new home, the statue means many things. To some, it means hope; to others, escape from hardship. To everyone, in some way, the statue means freedom.

Most people first see the Statue of Liberty from far away. It isn't until they get close that they spy the poem on the pedestal. Called "The New Colossus," this poem was written by Emma Lazarus in 1883.

Lazarus lived in a time when women were not treated equally. Raised in New York City, freedom was very important to her. Born Jewish-American, she was bothered by how the world's Jews were treated. She wanted to help. She visited temporary shelters overseas. She spoke at political gatherings. No matter what she tried, it wasn't enough—until she was asked to write the most important poem of her life. This poem helped raise money to build the Statue of Liberty's pedestal.

Lazarus was excited. Her poem was a message of freedom, not just for the Jews, but for the entire country. The poem's final lines read:

> Give me your tired, your poor,
> Your huddled masses yearning to breathe free,
> The wretched refuse of your teeming shore.
> Send these, the homeless, tempest-tost to me,
> I lift my lamp beside the golden door!

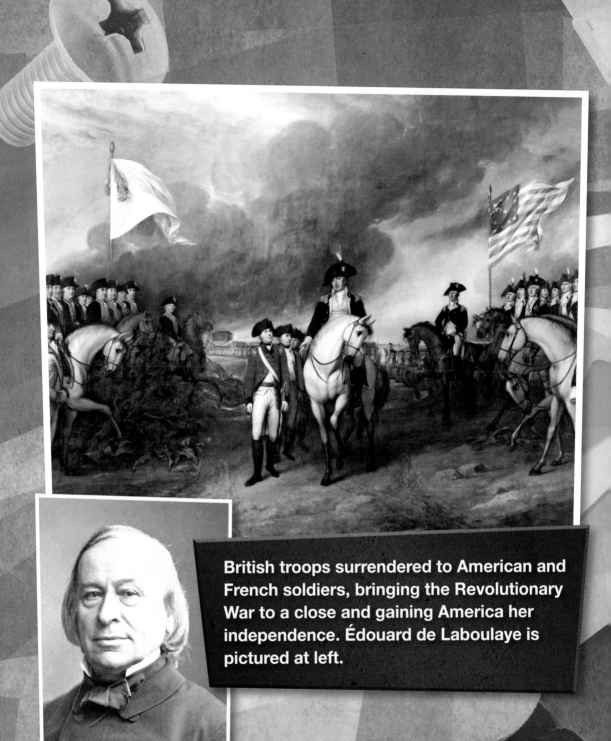

British troops surrendered to American and French soldiers, bringing the Revolutionary War to a close and gaining America her independence. Édouard de Laboulaye is pictured at left.

Friendship and Freedom

Building the Statue of Liberty was an enormous job. It took money and time. It required a lot of people and tools. It began, however, with a dinner party.

Édouard de Laboulaye (LAB-yoo-lay), a French lawyer, teacher, and historian, worried about his country. He wished it had the same freedoms as America, and he thought he could make a difference. He became an expert on the U.S. Constitution. This is the document that decides how the U.S. government is allowed to operate. It also guarantees American citizens certain rights and freedoms.[1] Laboulaye wanted the same for France. He looked to the past for the answers.

The United States of America wasn't always a single country. It began when people from Europe crossed the sea and set up camp. These camps became colonies, and England claimed 13 of these colonies for its own. Over time, the colonists decided they wanted to rule themselves. In 1776, they announced they wanted independence. It took a violent Revolutionary War, but they got it. They didn't have to fight alone. The British colonists were joined by fighters from France. They fought side by side against England. When the war ended, France and the United States stayed friends.

In Laboulaye's day, an emperor, Napoleon III, ruled France. Laboulaye thought his fellow Frenchmen would be happier if they lived under a democratic government. He also did not believe another war was needed to change politics. What did he need? He needed a dinner party—but not just any dinner party.

Frédéric-Auguste Bartholdi

In 1865, Laboulaye got together with a few of his friends who shared his dreams and worries. He asked, *Why not build a monument for their friends, the Americans, who were living out the ideals of liberty and equality?*

Frédéric-Auguste Bartholdi (bar-TOHL-dee) was sitting across from him. He liked this grand idea.

Bartholdi was a sculptor and a traveler. In Egypt, he saw the Sphinx and the pyramids. In Alexandria, he saw the 400-foot-tall lighthouse that had been standing for one thousand years. He saw the Colossus of Rhodes, a 110-foot-tall bronze figure of the sun god Helios.[2] These buildings inspired him. Bartholdi wanted to build something big that others would travel to see. He agreed to help Laboulaye make his dream come true. Together, they would create a statue.

Even though the two became a team, these men actually spent a lot of time apart. While Laboulaye stayed in France, Bartholdi took the

It took 31 tons of copper, 125 tons of iron, $870,000, and hundreds of engineers, craftsmen, and workers toiling for 21 years to turn the Statue of Liberty from an idea into an icon.

Bartholdi's colossal "Lion of Belfort" stands guard at Belfort Castle in France. This sandstone sculpture is 72 feet long and 36 feet high.

steamship *Péreire* to the United States. In 1871, he spent five months bouncing from coast to coast. Everywhere he went, he talked about liberty to anybody who would listen. In order for this plan to work, the costs had to be split between the two countries. France would provide the statue. America would provide the pedestal on which the statue would stand.[3]

The beginning of this dream was slow. Bartholdi had trouble getting America to agree to the costs. An immense challenge waited for him and Laboulaye. But they were determined to make their dream a reality.

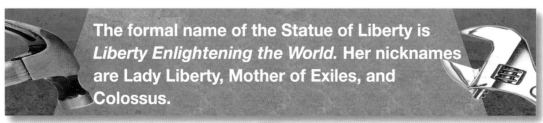

The formal name of the Statue of Liberty is *Liberty Enlightening the World*. Her nicknames are Lady Liberty, Mother of Exiles, and Colossus.

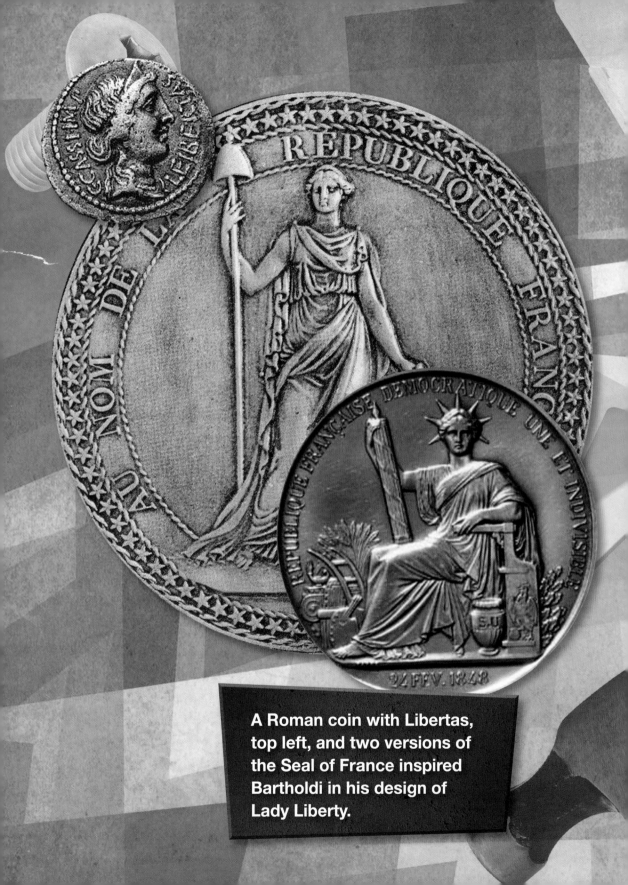

A Roman coin with Libertas, top left, and two versions of the Seal of France inspired Bartholdi in his design of Lady Liberty.

Designing Lady Liberty

Never before had anyone attempted to build a sculpture as large as Lady Liberty. Bartholdi planned for her to stand 151 feet tall. That's about 15 stories high!

There was much to do before the building could even begin. First, what would the statue look like? Bartholdi got his inspiration from the Roman Goddess of Freedom, Libertas. He studied Roman coins. He looked at French seals. On both, Libertas was shown as a woman in a long robe and a spiked crown. This was what Bartholdi wanted his statue to look like. Next, he needed to decide what to call her.

Bartholdi admired a painting by French artist Ange-Louis Janet-Lange. It was called *France Enlightening the World*. It portrays Liberty as a woman holding up a torch in one hand and a set of scales in the other. Because he liked the message in it, Bartholdi named the statue *Liberty Enlightening the World*.[1]

Now that he had his statue's name, it was time to make it real. Bartholdi started by making small models out of plaster and clay. With the help of his team in Paris, the models got larger and larger. The first model was four feet tall. The second was over nine feet tall. The third was 38 feet tall. The fourth and final model was 151 feet tall, the eventual size of the actual statue. These models helped Bartholdi and his team make sure his design would work. At one point, they realized they needed to bring Liberty's right arm closer to her head. This would help the arm to be better supported. It would also be safer from wind.

Bartholdi finished with his models and began creating the statue's actual parts. He did it a few pieces at a time, beginning with the head

Bartholdi is pictured in his studio in Paris with a model of the Statue of Liberty at center.

and right hand. Each plaster section was fitted with a wooden mold. The molds were used to shape Liberty's "skin."

Bartholdi wanted to use copper, but then he had a much more exciting idea. In order to make the statue easier to see from far away, he wanted to make her skin gold! He was disappointed. It was hard enough getting the money for the statue without using gold.

Bartholdi chose to work with copper. This fairly inexpensive metal bends easily and lasts a long time. Soon, many sheets of copper were arriving at his studio. In the end, 62,000 pounds of copper were used to make the statue. The thickness of each copper sheet was ³⁄₃₂ inch. This is about the width of two stacked pennies.[2]

Putting the copper skin on Liberty was tricky. Bartholdi hired the metalworking firm of Gaget, Gauthier & Company in Paris to

Coppersmiths work with sheets of copper and wooden molds.

help. This firm also provided a larger workspace for building the statue.

According to tradition, Bartholdi used his mother, Charlotte, as the model for Liberty's face.

Workers involved in the project used a technique called repoussé. They carefully hammered the sheets of copper into the wooden molds. From beginning to end, more than 300 kinds of hammers were used. When the copper was separated from the mold, it was in the proper shape to become part of the face, hand, or foot. Eventually, the hundreds of shaped copper sheets would be fitted together like a giant puzzle. More than 600,000 rivets would hold the pieces in place.

In order to stand up, the hollow statue had to have a skeleton inside to support her. Who did Laboulaye and Bartholdi bring in to help with this? The man who would build the Eiffel Tower, Alexandre-Gustave Eiffel himself!

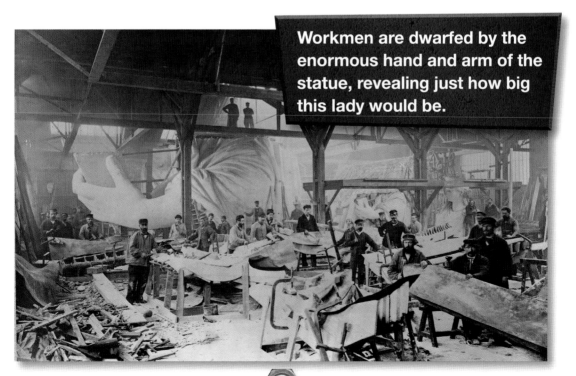

Workmen are dwarfed by the enormous hand and arm of the statue, revealing just how big this lady would be.

Alexandre-Gustave Eiffel

As Liberty's head and arm await placement upon the torso, her skeleton can be seen rising high above the Paris skyline.

Inside the Lady From Head to Toe

When Eiffel was first brought in to help with the Statue of Liberty, he hadn't yet built the Eiffel Tower in Paris. He was best known for building iron railway bridges. His largest bridge was more than 1,800 feet long. Designing this bridge prepared him for designing the framework that would support Liberty's copper skin.

While the many coppersmiths continued working on Liberty's skin, Eiffel focused on her skeleton. He began with a tall pylon of four vertical iron beams connected by cross braces.[1] From this central pylon he extended a web of secondary beams that stretched toward the copper skin, but did not actually touch it. Iron and copper expand and contract at different rates when the temperature changes. Eiffel knew that the statue would have to endure heavy winds. He wanted to protect it from that. To do this, he knew the skeleton (iron beams) and the skin (copper sheeting) could not be secured to each other.[2] Instead, flat iron bars were attached to the end of each beam. These were attached to copper brackets connected to the skin. The thin iron bars acted like springs. Once put together, the copper skin would be able to bob around in the wind and withstand changing temperatures.[3]

Another problem faced Eiffel. Direct contact between the statue's skeleton and the statue's skin could cause the iron to rust. To solve this problem, workmen slid fabrics between the iron supports and the copper sheets. The fabrics were covered with red lead or were made of asbestos. They acted as insulators and protected against rust.

Up, up, up went the skeleton until it could be seen from all over Paris. When this frame was completed in 1881, it was time to bring out

The torch in Philadelphia

Liberty's skin. The first part of Liberty's body to be attached to the frame was her big toe.

By this time, two sections of Lady Liberty had been completed. They were on public display for a number of years. In 1876, her right hand holding the torch sailed to the United States. The torch, called the Flame of Freedom, stood for the ideas of enlightenment and liberty. Sending the hand and torch to the United States helped raise money for the project. For a small fee, visitors could sit in the palm of Lady Liberty's hand and have their picture taken. They could also pay to climb up the stairs to the torch or to the windows in the crown. When Liberty's skeleton was finished, the head and hand were brought back to the workshop to be connected to the frame.

In 1884, the statue was finished. On July 4, French leaders invited a few American leaders to come and see it in Paris. Hundreds of French people showed up as well. They wanted to see Lady Liberty before she took her great trip across the ocean to New York.

Liberty Enlightening the World was finished. Paris had played its part in this grand masterpiece. Now it was America's turn.

In 50-mile-per-hour winds, the statue may sway up to three inches, while the torch may sway up to six inches.[4]

The completed statue, visible from all over Paris, would soon be ready to sail to her permanent home in New York.

The cornerstone for the pedestal was laid on a rainy day in August 1884. This was the first step in building the enormous stand from which Lady Liberty would shine her light.

Richard Morris Hunt

A Place For the Lady to Stand

What the statue looked like and what to call it were very important decisions. What was almost as important? Where to put it! The statue almost ended up in Fairmount Park, Philadelphia. The torch was displayed there for the World's Fair. Fairgoers paid a fee to climb into the torch. The interest in the torch was so overwhelming, Bartholdi almost decided to keep the entire statue in Philadelphia. Boston, too, wanted the statue for itself. In the end, Bartholdi's first choice of New York won.

There was a small piece of land in New York Harbor called Bedloe's Island. In the future, it would be called Liberty Island. The only structure on the island was the little-used military fort, Fort Wood. It was shaped like a star with eleven points. Bartholdi decided the fort would make the perfect pedestal for his huge statue. From this spot, Lady Liberty could be seen from the city and from the mainland. It could be seen from the harbor and from every ship that brought immigrants from the Old World to the New. Bartholdi also wanted the statue on this island so that it could be used as a lighthouse.[1]

Once he found the right spot for Lady Liberty, Bartholdi needed to find the person to build the pedestal. He brought in American architect Richard Morris Hunt. Hunt's design called for the use of a relatively new material: concrete. Concrete is a mixture of sand, cement, and small stones. Combined with steel bars, it is even stronger than stone. When finished, Hunt's pedestal would weigh more than 27,000 tons. This would make it the largest concrete structure of the time.

Chapter 4

If Lady Liberty were a real person, she would wear a size 879 shoe! The length of her right arm is 42 feet, her forefinger is 8 feet, her nose is 4½ feet, and the width of her mouth is 3 feet. Her total weight is 450,000 pounds. The climb to her crown from the bottom of the pedestal is 354 steps. That's one super lady![3]

Joseph Pulitzer

Charles P. Stone, an American engineer, would oversee construction of the pedestal. His first step was to dig a hole 17 feet deep in the center of Fort Wood. The hole was 91 square feet at the bottom and 65 square feet at the top. Once the hole was dug, workers poured concrete in it. This was where the 89-foot-high pedestal would stand. This base would be concrete also, but it would be faced with granite blocks. A steam-powered crane was used to hoist the materials for the pedestal.

In 1877, the American Committee was formed to oversee the fundraising. This committee sold small bronze versions of the statue. It also hosted fairs, concerts, and theatrical events. Donations were often small and never enough. By the time Bartholdi completed the statue in Paris, only $16,000 of the $300,000 needed for the base had been raised.

In 1885, Joseph Pulitzer came to the rescue. Pulitzer was from Hungary and had been very poor when he arrived in America. Now he was the successful owner and publisher of the New York newspaper *The World*. For five months, Pulitzer wrote editorials for his newspaper urging people to help fund the pedestal. He wrote how disappointed he was that the American people were not donating more. He knew that not everyone could afford it, so he promised to publish the name of every person who made a donation, no matter how small.

Soon pennies, nickels, dimes, and dollars came in from all around the country. Much of it came from children. A class of kindergarten students in Iowa sent $1.35. An office boy in New York sent in a nickel. More than 120,000 people responded to Pulitzer's request for funds. By the end of Pulitzer's effort, more than $100,000 had been raised. Finally, this would be enough to finish the pedestal. Pulitzer was true to his word, too. He published the name of every donor.[2]

A GIFT
FROM
THE PEOPLE OF THE REPUBLIC OF FRANCE
TO THE PEOPLE OF THE UNITED STATES.
THIS STATUE
OF
LIBERTY ENLIGHTENING THE WORLD
COMMEMORATES THE ALLIANCE OF THE TWO NATIONS
IN ACHIEVING THE INDEPENDENCE
OF THE
UNITED STATES OF AMERICA,
AND
ATTESTS THEIR ABIDING FRIENDSHIP.
AUGUSTE BARTHOLDI.
SCULPTOR.
INAUGURATED
OCTOBER 28TH 1886

The face is uncrated and both the inside and out are inspected. Above: The dedication plaque.

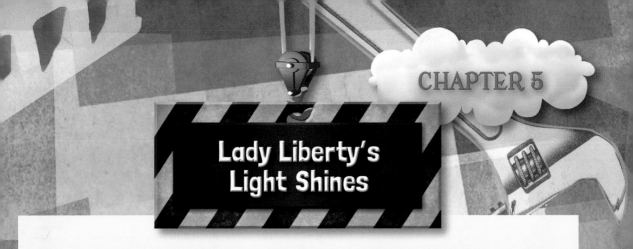

Lady Liberty's Light Shines

The time was at hand! It was 1885 and Lady Liberty was finished. She was ready to be taken across the ocean to the United States. Very carefully, the statue was taken apart and packed into 214 crates. Every piece had to be labeled so that the workers in New York would know how to put her together again. The crates were loaded onto 70 train cars. The train took them to the harbor to be loaded onto the French naval ship *Isère*. Getting them on board took 17 days.

The voyage from France to New York took nearly a month. The seas were stormy and the ship nearly sank many times, but *Isère* made it! On June 19, Lady Liberty arrived at Bedloe's Island. She was ready to be put right where she belonged. Unfortunately, America wasn't ready for her. The pedestal was still not finished.

The 214 crates were unloaded but not unpacked. With the pedestal unfinished due to lack of funding, Lady Liberty would end up staying in boxes for nearly a year.

Finally, in 1886, the pedestal was done. It was time for Americans to meet Lady Liberty. First, the central pylon was anchored to steel beams built right into the concrete walls of the pedestal. This would help keep the statue from falling over. Once this central pylon was in place, a steam-powered crane was mounted on its top. This crane helped pull the rest of Liberty's frame and copper plates into place. To put on the copper skin, riveters climbed up the iron framework inside the statue, going higher and higher as they worked. When the sun shone on the copper, it got extremely hot inside.

Unveiling the Statue of Liberty **painting by Edward Moran**

No scaffolding was used outside the statue. The workers had to dangle on ropes up to 305 feet above the ground. It was dangerous work, but the job was completed without any serious accidents.[1]

On October 25, 1886, the last rivet was driven into place. Imagine the shiny giant complete at last, her torch reaching high in the sky over the harbor!

The shiny copper did not stay reddish gold for long, though. When copper meets damp air, a protective green coating forms over the metal. By 1915, Lady Liberty was completely green. She remains so today.

A dedication and unveiling ceremony was held on October 28, 1886. President Grover Cleveland was head of the ceremony. Auguste Bartholdi sailed from France to New York for the event. Many other French and American representatives attended, including architect Hunt and metal expert Gaget, and engineer Stone. One person not present for the ceremony was Laboulaye. He had died in 1883 and never saw his dream made a reality.

The celebration began with a five-mile parade through lower Manhattan. Millions of people lined the parade route, waving both American and French flags. After the parade ended, dignitaries were taken by boat to Bedloe's Island. Crowds of spectators lined the harbor's shore. More than 300 ships and boats crowded the waters of the bay.[1]

Bartholdi had a very important job. He was supposed to climb the 354 stairs to the crown of the statue. At a special signal, he was to drop the French flag that covered Liberty's face. Before he could, chairman of the American Committee, Senator William M. Evarts, started his speech. While Senator Evarts was talking, he paused to take a breath. During the pause, the signal was accidentally given to Bartholdi. Bartholdi released the flag.

For the first time, Americans saw the face of Lady Liberty. The crowd went wild! Whistles blew, cannons roared, guns fired salutes, and brass bands played. The thunderous noise went on and on. In the end, Senator Evarts gave up on his speech and sat down.[2]

When the crowd finally quieted, President Cleveland took the stand. On behalf of the American people, he accepted the gift of *Liberty Enlightening the World* from the country's friend and ally, France. The colossal symbol of freedom and friendship between the two nations was home at last.

On November 1, 1886, lights inside the torch were switched on and Liberty became a lighthouse which could be seen for 24 miles. It remained a beacon for ships at sea until 1902.

Lady Liberty today

1865 At a dinner party in Paris, Édouard de Laboulaye presents his idea for the Statue of Liberty.

1871 Frédéric-Auguste Bartholdi, the statue's designer, travels to America.

1875 The Franco-American Union is formed and becomes the official sponsor for the project. Work begins on the statue.

1876 The completed right hand and torch go on display in Philadelphia and then in New York. Two years later, Liberty's head goes on display in Paris.

1877 President Ulysses S. Grant signs a resolution accepting the gift of the Statue of Liberty. The American Committee is formed to sponsor construction of the pedestal.

1880 Alexandre-Gustave Eiffel designs the frame for the statue.

1881 Noted American architect Richard Morris Hunt is commissioned to design the pedestal. American engineer Charles P. Stone is hired to oversee its construction.

1884 The Statue of Liberty is officially presented to the United States during a ceremony held in Paris on July 4.

1885 Joseph Pulitzer begins his fundraising campaign. The statue is disassembled, packed into 214 crates, and shipped to the United States.

1886 The pedestal is completed. Workers put the statue together on the pedestal. On October 28, *Liberty Enlightening the World* is unveiled in one of the biggest celebrations New York has ever seen. President Grover Cleveland accepts the statue on behalf of the American people.

1903 "The New Colossus" by Emma Lazarus is inscribed on a bronze tablet and placed on Liberty's pedestal.

1907 The first elevator is installed inside the pedestal.

1916 The torch is permanently closed to the public.

1924 The Statue of Liberty is declared a national monument.

1936 Franklin Roosevelt rededicates the statue on October 28.

1956 Bedloe's Island is renamed Liberty Island.

1986 One hundred years after her unveiling, the Statue of Liberty receives a complete restoration. Changes include replacing the torch with one covered in gold leaf. President Ronald Reagan rededicates the statue.

2001 After the terrorist attacks on U.S. soil on September 11, the Statue of Liberty is closed to the public.

2009 The Statue of Liberty is fully reopened, with visitors allowed to go all the way to the crown once again.

2012 Damage from Hurricane Sandy closes the statue again.

2013 On July 4, the statue reopens to the public. Repairs to the island continue.

2015 The Statue of Liberty 5K, hosted by the Special Olympics, is held on Liberty Island on the morning of July 4.

2016 A Broadway show about the Statue of Liberty, called *Liberty: A Monumental New Musical*, opened in New York on July 4.

The Statue of Liberty greets the *Queen Mary II.*

Chapter 1. Friendship and Freedom

1. Barry Moreno, *Images of America: The Statue of Liberty* (San Francisco: Arcadia Publishing, 2004), p. 10.
2. Yasmin Sabina Khan, *Enlightening the World: The Creation of the Statue of Liberty.* (Ithaca, NY: Cornell University Press, 2010), pp. 100–101.
3. Elizabeth Mitchell, *Liberty's Torch* (New York: Atlantic Monthly Press, 2014), p. 110.

Chapter 2. Designing Lady Liberty

1. Edward Berenson, *The Statue of Liberty: A Transatlantic Story* (New Haven: Yale University Press, 2012), p. 13.
2. "Statue Statistics," *National Park Service: Statue of Liberty,* accessed April 7, 2015, http://www.nps.gov/stli/learn/historyculture/statue-statistics.htm

Chapter 3. Inside the Lady from Head to Toe

1. Edward Berenson, *The Statue of Liberty: A Transatlantic Story* (New Haven: Yale University Press, 2012), p. 59.
2. Elizabeth Mitchell, *Liberty's Torch* (New York: Atlantic Monthly Press, 2014), p.p. 148, 150.
3. Berenson, p. 59–60.
4. "Statue Statistics," *National Park Service: Statue of Liberty,* accessed April 7/2015, http://www.nps.gov/stli/learn/historyculture/statue-statistics.htm

Chapter 4. A Place for the Lady to Stand

1. Edward Berenson, *The Statue of Liberty: A Transatlantic Story* (New Haven: Yale University Press, 2012), p. 31.
2. "Statue of Liberty," *New York Architecture,* accessed April 15, 2015, http://www.nyc-architecture.com/LM/LM002-STATUEOFLIBERTY.htm
3. Yasmin Sabina Khan, *Enlightening the World: The Creation of the Statue of Liberty.* (Ithaca, NY: Cornell University Press, 2010), p. 172.

Chapter 5. Lady Liberty's Light Shines

1. Elizabeth Mitchell, *Liberty's Torch* (New York: Atlantic Monthly Press, 2014), pp. 241–242.
2. Yasmin Sabina Khan, *Enlightening the World: The Creation of the Statue of Liberty.* (Ithaca, NY: Cornell University Press, 2010), pp. 178–179.

Books

Behrens, Janice. *What Is the Statue of Liberty?* New York: Scholastic, Inc., 2009.

Glaser, Linda. *Emma's Poem: The Voice of the Statue of Liberty.* New York: Houghton Mifflin Books for Children, 2013.

Holub, Joan. *What Is the Statue of Liberty?* New York: Grosset & Dunlap, The Penguin Group, 2014.

Malam, John. *You Wouldn't Want to Be a Worker on the Statue of Liberty!* New York: Franklin Watts, An Imprint of Scholastic Inc., 2009.

Rappaport, Doreen. *Lady Liberty: A Biography.* Cambridge, MA: Candlewick Press, 2008.

Works Consulted

Berenson, Edward. *The Statue of Liberty: A Transatlantic Story* (Icons of America series). New Haven: Yale University Press, 2012.

Burchard, Sue. *The Statue of Liberty: Birth to Rebirth.* New York: Harcourt Brace Jovanovich, Publishers, 1985.

Khan, Yasmin Sabina. *Enlightening the World: The Creation of the Statue of Liberty.* Ithaca, NY: Cornell University Press, 2010.

Mitchell, Elizabeth. *Liberty's Torch.* New York: Atlantic Monthly Press, 2014.

Moreno, Barry. *Images of America: The Statue of Liberty.* San Francisco: Arcadia Publishing, 2004.

On the Internet

Engineering Facts—Statue of Liberty: Science Kids
http://www.sciencekids.co.nz/sciencefacts/engineering/statueofliberty.html

Great Buildings: Statue of Liberty
http://www.greatbuildings.com/buildings/Statue_of_Liberty.html

The Light of Liberty: National Geographic Kids
http://kids.nationalgeographic.com/kids/stories/history/statue-of-liberty/

National Park Service: Statue of Liberty
http://www.nps.gov/stli/learn/historyculture/statue-statistics.htm

New York Architecture: Statue of Liberty
http://www.nyc-architecture.com/LM/LM002-STATUEOFLIBERTY.htm

The Statue of Liberty and Ellis Island
https://www.libertyellisfoundation.org/

Statue of Liberty Fun Facts—Kidzworld
http://www.kidzworld.com/article/2512-statue-of-liberty-fun-facts

ally (AL-eye)—A friend or helper.

architect (AR-kih-tekt)—A person who designs buildings.

asbestos (as-BES-tus)—A soft mineral that does not burn, once used for making fireproof materials and now known to cause cancer.

colony (KAH-luh-nee)—A group of people who settle in a distant land but who remain under the rule of another country.

colossus (kuh-LAH-sus)—Gigantic, much larger than life-size; a gigantic structure.

concrete (KON-kreet)—A strong building material formed by mixing cement, sand, gravel, and water and allowing it to cure.

constitution (kon-stih-TOO-shun)—The fundamental principles of a government or nation. The U.S. Constitution guarantees certain basic rights and freedoms for its citizens.

copper (KAH-pur)—A reddish yellow metal that is soft and easy to shape.

coppersmith (KAH-pur-smith)—A person who works or makes things with copper.

democracy (deh-MAH-kruh-see)—Governed by the people (through voting) instead of by a ruler.

emperor (EM-per-or)—A person who has supreme rule over a country or empire.

engineer (en-jeh-NEER)—Someone who puts scientific knowledge to practical use, such as in designing and constructing roads, bridges, and buildings.

equality (ee-KWAH-luh-tee)—The state of being equal, such as all people enjoying the same rights and privileges.

exile (EK-zyl)—A person who leaves his or her home country to live in another.

expand (ek-SPAND)—To spread out or make larger.

granite (GRAN-it)—A very hard rock that varies in color from gray to pink.

immigrant (IM-uh-grunt)—A person who arrives in a country from another country to live permanently.

insulator (IN-suh-lay-ter)—Anything that keeps electricity, heat, or sound from passing from one object to another.

masterpiece (MAS-ter-pees)—The greatest work made by a person or group.

mold—A shape made out of wood or plaster from which additional identical shapes can be made.

pedestal (PED-es-tel)—The base on which a statue stands.

plaster (PLAS-ter)—A white powder made from the mineral gypsum. When combined with water, it can be used for molding and sculpting.

red lead (RED LED)—A type of paint that coats and protects metal.

repoussé (ruh-POO-say)—The process of hammering metal on the underside of a mold to give it a desired shape.

rivet (RIV-it)—A short metal rod with a head on one end used to fasten pieces of metal together. The other end is hammered to make a second head so that the metal pieces are held tight.

seal—Melted wax used to close a letter, usually stamped with a symbol.

sculptor (SKULP-ter)—An artist who models or carves statues out of stone, wood, or other materials.

PHOTO CREDITS: pp. 4, 13—Library of Congress; pp. 12, 14, 18, 21—NPS.gov; All other photos—Public Domain. Every measure has been taken to find all copyright holders of material used in this book. In the event any mistakes or omissions have happened within, attempts to correct them will be made in future editions of the book.

Index